Story by Gail Tuchman

Illustrations by Mary Ann Lloyd

HARCOURT BRACE & COMPANY

Orlando Atlanta Austin Boston San Francisco Chicago Dallas New York
Toronto London

 Can you find this hen? 

Can you find this mouse?

Can you find this cow?

Can you find this pig? 

Can you find this sheep? 

Can you find this horse? 

Did you find them?